Hot Foot Teddy

The True Story of Smokey Bear

By
Sue Houser

ONLY YOU

OFFICIAL LICENSEE · SMOKEY

COOPERATIVE FOREST FIRE PREVENTION PROGRAM

16 USC 580 P-4

M.T. Publishing Company, Inc.
P.O. Box 6802
Evansville, Indiana 47719-6802
www.mtpublishing.com

Graphic Design:
Alena L. Kiefer

Library of Congress Control Number:
2006906877

ISBN: 1-932439-56-0

First Printing: March 2007
Second Printing: September 2007
Third Printing: June 2008

Printed in the
United States of America

To my writing group
Lynne, Nadine, Rob
and Wendy

and

Roger.

A portion of the proceeds from the sale of this book go toward the Smokey Bear Forest Fire Prevention program.

Trapped In the Fire

Flames leaped from ridge to ridge. Trees sizzled in red and yellow blazes. Underbrush crackled, branches snapped, and trees fell. Thick, dark smoke choked the air. Capitan Mountain in southern New Mexico was on fire.

Weary firefighters battled the stubborn flames all night. As they trudged back toward camp, they heard whining sounds. A bear cub, wailing as if he were a lost child, emerged from the underbrush. When the cub saw the tired dirty crew, the howling stopped.

From a distance, the cub and the men studied each other. The little bear appeared to be alone. Should the firefighters rescue the small helpless cub? What if the mother was nearby?

Capitan Gap Fire ruined the forest.

Finally, Game Warden Speed Simmons and his relief crew arrived. "We've got work to do, fellas. Don't have time to be making friends with a bear cub," said Simmons. "Anyway, this is a game refuge. The rules say to leave wild animals where they're found."

Simmons and his crew hiked on toward the towering flames. The cub trailed after them, then wandered off.

That evening while sitting around the crackling fire, the story of the bear cub was told. "He was a cute little thing—chocolate colored—couple months old, maybe—stopped crying when he saw us—wasn't afraid, just curious."

Ray Bell, the Game and Fish Chief Field Officer, listened closely. He worried that the mother might have died in the fire, leaving the cub to live on his own. Bell waited awhile before he spoke. "I know it is against the rules, but if you see this little bear again, bring him in. He may be too young to survive on his own."

The next day, fierce winds fueled the fire. Trees exploded. Boulders burst open from the heat. Speed Simmons' crew of 24 men was trapped.

Fire blew up around them. Engulfed in smoke, Simmons ordered the men to take refuge in a small rocky ravine. They soaked handkerchiefs with water from canteens. In order to breathe, they held the wet cloths over their faces. Squeezed together, lying face down, the men flattened themselves against the rocks, and waited. They only moved to swat flying sparks that landed on each other's hair and clothing.

But Simmons and his crew were not the only ones trapped in the fire. Clinging to a charred tree at the edge of the fire was a tiny bear cub weighing about five pounds. His hair was singed. His feet were burned and bloody. His bottom was scorched.

1944 Drawing by Albert Staehle.

SMOKEY SAYS –
Care will prevent
9 out of 10 forest fires!

Rescued

Finally, the winds died and the smoke cleared. Simmons' crew went to work setting a new fire line. Faint, pitiful cries from a burned tree caught their attention. One man climbed the tree and slipped the whimpering cub inside his jacket. Back on the ground, the firefighter cradled the cub in his arms and gave him water. Another firefighter had a first-aid kit. He bandaged the cub's tiny paws and treated his burned behind. "He's burned pretty bad. We can't leave him here. Let's take him back to camp."

The little bear snuggled on the shoulder of the firefighter who had rescued him. However, if anyone else tried to pet him, he snapped and growled at them.

The men felt a bond of friendship with this little cub. After all, he had been trapped in the fire with them. They tried to comfort him. The camp cook, Ray Taylor, poured the bear a spoonful of milk. The tiny bear promptly bit the spoon. One firefighter waved a can of milk under the bear's nose. The bear grabbed it and chugged down the whole can. He stuck his bandaged paw into a jar of grape jam and slowly licked it off. However, the injured bear continued to whimper and whine.

Ray Bell stood nearby. "It's a good thing you brought this cub in. He couldn't have survived in the mountains." Then, Bell threw his sleeping bag on the ground. He had not slept in several days and fell into a deep sleep.

The other firefighters, however, had bigger problems than a hungry, unhappy cub. "We have to get back to fighting the fire. We don't have time to take care of this bear," said one.

When Ross Flatley offered to take the bear to his ranch nearby, no one objected. So Flatley drove home that night with the whimpering cub.

6

17,000 acres of forest was destroyed on
May 11, 1950

When Ray Bell woke the next morning, he announced, "I'll take the bear home with me." Bell often took injured animals home. In fact, his family had cared for deer, birds, beaver, and even bears, but not one this small.

Flatley, a part-time Forest Service employee, explained to Bell that the bear was at his ranch and offered to drive Bell over to get the bear. Bell jumped into Flatley's two-door Chevrolet coupe, and they headed for the ranch.

Mrs. Flatley met Bell in the front yard. She handed Bell a cardboard box with a ball of fur curled up inside. "You can have him, Ray," she teased. "We didn't get a wink of sleep last night. This little rascal yelped and carried on so that the dogs and coyotes howled all night."

The ride back to camp was bumpy. The little cub let out piercing screams each time he was bounced on the rough roads. "Maybe this isn't such a good idea," Bell muttered to himself.

But there was another reason Bell was interested in this bear's recovery. The Forest Service had chosen a Smokey Bear fire prevention symbol in 1944. Now, the Game and Fish Department had a live bear to go with the symbol if the bear could be saved and if both agencies agreed.

Before leaving the area, Ray Bell surveyed the damage of the Capitan Gap fire.

He then flew his plane over the Capitan Mountains. Looking over the 17,000 acres of blackened forest, he shook his head. It would take 400 years to restore the mountain that had been destroyed in five days.

The First Flight

"The fire appears to be under control. We can go home," Bell said. He glanced toward the bear. "But first, I'm taking you to the doctor." So Bell and the little bear took off on a 50-minute flight to Santa Fe.

Their first stop was a visit to a veterinarian, Edwin Smith. Dr. Smith cleaned the cub's burns and applied antiseptic and soothing ointments. He wrapped the small paws in gauze and tape. The exam lasted 45 minutes as the vet checked for swollen glands, ticks, and lice. He then examined the cub's eyes and teeth. Surprisingly, the bear cooperated. "I would guess the bear to be about two and a half months old," said Dr. Smith. "He is pretty sick, and he is badly burned. But with good care, he'll soon be running around."

Following the exam, the sickly little bear curled back up in the cardboard box and went to sleep.

Dr. Smith and his staff tried to feed him, but when he was offered food, he turned his head away.

Ray Bell worried that the cub was not eating and getting weaker so he took home the little fur ball in the cardboard box.

A **Always break matches in two!**

B **Be sure fires are out cold!**

C **Crush all smokes dead!**

PLEASE!

Only you can **PREVENT FOREST FIRES**

Smokey's ABC's

10

A Foster Family: The Bells

Don Bell, a teenager, and five-year old Judy Bell were used to unusual houseguests. This one, they called "The Bear." (Later, he was known as "Smokey.") Bell's wife, Ruth, had nursed many sick animals and, right away, she made a mixture of milk, Pablum (baby cereal) and honey. Covering her index finger with the mixture, she rubbed the paste over the bear's teeth and back of his throat. He smacked his lips and swallowed. At night, Ruth set her alarm for every two hours to feed the helpless bruin. Sometimes, Don and Judy helped their mother, holding the bear while Ruth fed it.

Soon, the bear was stronger. He shared the laundry room with the Bell's cocker spaniel, Jet, where they ate puppy food together from the same dish. They rolled and tumbled and chased each other through the house. Jet even trained the bear to go to the bathroom on newspaper! When the bear was in the backyard and needed to use the bathroom, he scratched and whined until someone opened the door. Then, he ran to the newspaper and did his business.

One day, the little bear could not be found. The Bells searched the house. They combed the neighborhood. Where could he be? Finally, they returned home to find the napping bear. He had fallen into the washing machine and dozed off.

Once, when Bell and Smokey were playing in the backyard, the little bear nipped Bell. Bell backhanded him to teach him a lesson, as would a mother bear. The little bear had a temper tantrum. He raised his paws over his head and rolled on the ground, screaming and wailing.

When Bell took Smokey to meet his co-workers at the Game and Fish Department, Smokey tried to bite all of them. And, when Bell brought home another orphaned bear cub, Smokey snapped at him as well.

"The bear was not a family pet. He was a wild animal," recalled Don Bell, years later. "He bit Dad and me every time he could. I still have scars from his bites."

However, not once did Smokey attempt to bite Ruth, or Judy, or his playmate, Jet.

Earning His Keep

During this time, the U.S. Forest Service and the New Mexico Game and Fish Department made many phone calls and held many meetings.

"What should we do with this bear? He can't stay at the Bell house much longer."

"Could this little cub that survived a forest fire become the live counterpart to the Smokey Bear that advertised 'Only You Can Prevent Forest Fires'?"

"We already have a poster bear named Smokey Bear. Why do we need another bear?" the Forest Service officials wondered.

The New Mexico Game and Fish Department employees argued. "But we have a tiny cub that survived a forest fire. He could campaign for forest fire prevention—a living symbol."

And so the meetings and phone calls went back and forth, back and forth.

Dr. Smith bandages Hot Foot Teddy!

12

Meanwhile, the story of the rescued bear leaked to the press. On May 11, 1950, the Santa Fe New Mexican ran an article, "Teddy with a Hotfoot" with a photograph of Dr. Smith bandaging the cub's paws.

On May 14, 1950, the same newspaper printed a picture of the bear captioned, "I'm A Stranger Here, Myself," and said "Hot Foot Teddy" would be going home when his injuries healed.

However, the bear was now a celebrity and was called "Smokey." He even went on the radio. Elliott Barker, Ray Bell's boss at the Fish and Game Department, frequently did radio broadcasts about forest fires. Barker invited Smokey to join him in the studio. Barker smeared honey on the microphone and let the bear lick it off. Then he pulled the bear away causing him to scream for the honey. Barker announced that "Smokey Bear was live on the air."

When photographer Harold Walter came to take pictures, the bear wasn't interested. He ran around the house on bandaged paws, fussing and whining. Finally, Walter smeared honey on five-year-old Judy Bell's boot. The bear promptly licked it off. Walter dabbed honey on Judy's knee. The bear licked it off. Then, he smeared honey on Judy's chin, and the bear licked it off.

The popular bruin was invited to a number of schools. On his visits, he wore a leash held by Ray Bell. Bell didn't want to take a chance on Smokey biting someone.

Bell told the children about Smokey being rescued from the forest fire with burned paws and a scorched behind. "Don't play with matches," reminded Bell. "And, if you go on a picnic or go camping, make sure the campfire is out before you leave."

One meeting lasted too long, and Smokey was in a nasty mood. On the ride home, Smokey pounced on Bell who was driving. Bell shoved Smokey onto the lap of his co-worker, Lee Leavon. Smokey swatted Lee. Then, Lee pushed Smokey back onto Bell, and Smokey nipped Bell on the belly. "Ouch! You little rascal!" shouted Bell. Lee tried to grab the leash. Bell tried to keep his truck on the road. And the little fur ball wrestled with both men who continued on their way, laughing.

Perhaps Smokey's behavior was normal. His situation was not normal. After all, the little bear had lost his mother and had been burned in a forest fire. Now he was living in a strange, human world. Why wouldn't Smokey bite and howl?

Judy Bell and Smokey Bear

Saying Good-Bye

Finally, after all the meetings and phone calls, the Forest Service sent word: Smokey Bear was invited to live at the National Zoo in Washington, D.C. Both the live Smokey Bear and the poster Smokey Bear would be symbols of forest fire prevention.

Now they had to make plans to fly Smokey to Washington, D.C. There were more meetings and more phone calls.

Homer Pickens of the New Mexico Game and Fish Department was one of the men selected to accompany Smokey. Pickens wanted to become better acquainted with Smokey before the trip. So Ruth, Don, and Judy Bell took Smokey, in his cage, to the Pickens' home.

Confined to the Pickens' fenced yard, Smokey acted like a puppy. He played with their dog, Sarge. He romped with two mountain lion kittens that were staying at the Pickens' house. He climbed the apple tree. But because of Smokey's reputation, the family stayed away from him. Pickens, himself, always wore buckskin gloves when handling the feisty little bear.

On June 26, 1950, the Boy Scouts and Girl Scouts threw a going-away party for Smokey in Santa Fe. Over 150 people attended, including the mayor of Santa Fe and others with official titles. However, the speakers could not compete with Smokey. When he tried to climb out of his glass cage, the crowd roared with laughter. Smokey stole the show!

Ray Bell gives Smokey Bear a
drink of soda.

16

Before the sun came up the next morning, a crowd gathered at the airstrip. "Look at Smokey's plane!" said a young girl. Sitting on the runway was a Piper Cub. The name "Smokey" and a small bear with its arm in a sling had been painted on the side of the airplane.

This was a special flight, and the men making the trip dressed up for the occasion. Homer Pickens wore a Game and Fish Department uniform with a necktie; Kester Flock wore a tie with his green Forest Service uniform; and the pilot, Frank Hines, also wore a tie.

A few photographs were taken. Then the pilot announced, "Time to board!" The plane taxied down the runway and lifted into the air.

On the ground, children waved. "Goodbye, Smokey. I'll come and visit you at the zoo."

Smokey Bear and Homer Pickens with Piper Cub airplane.

Smokey Bear and Homer Pickens
Inside the Plane

18

In The Air, Again

Smokey's three-day, cross-country trip included stops for refueling and bathroom breaks. Smokey rode in a cage and slept most of the time except during take-offs and landings. Then he scratched his ears and whined.

Before they reached Amarillo, pilot Hines called the airport for instructions to land. When asked how many people were aboard the plane, the pilot replied, "Three men and a bear." The tower radioed back with a stern warning: "Don't joke about air traffic. This is a serious matter."

Finally, they were cleared for landing. Among their greeters were two surprised air safety officers. "You weren't joking. You do have a bear passenger!"

In St. Louis, Chase Manhattan Bank people sent a black limousine to pick up the men. Smokey spent the night in the basement of the reptile cage at the Forest Park Zoo. When they stopped in Cincinnati, this time it was Smokey that was picked up by a limo. The men were taken to a reception.

When they landed in Amarillo, Tulsa, St. Louis, Cincinnati, and Baltimore, hundreds of people met them. News reporters and photographers pushed their way to the front of the crowd, hoping to get a Smokey Bear story for their home town news.

Smokey Bear poster and Smokey Bear Cub

PLEASE FOLKS, be extra careful this year!

Remember - Only you can PREVENT FOREST FIRES

Welcome, Smokey Bear!

Smokey arrived in Washington, D.C. on June 29, 1950. It was reported that President Truman granted permission for the little aircraft to fly over the White House. Twenty-seven commercial airliners were held in circling patterns while Smokey Bear's little plane landed.

Over two hundred Smokey admirers and government officials waited in pouring rain to greet them. Smokey put on a show for his waiting fans. He reared up on his hind legs and pranced through water puddles along the tarmac, catching raindrops on his snout. Inside the terminal, TWA hostesses rewarded Smokey with a warm bottle of milk while Pickens held the leash. The following day, a newspaper headline read, "First Washington Lunch."

TWA hostesses feed Smokey lunch.

21

"What was it like to fly with a bear?" news reporters wanted to know.

"Oh, Smokey enjoyed the flight. He slept most of the time," Pickens replied. Just then, Smokey swatted Pickens. The questions continued, and Pickens dodged Smokey's lunges and nips. The audience was thrilled.

Smokey was then transported to the National Zoo by motorcade. His official registration form at the zoo read:

General classification: Carnivora-Ursidae
Specific name: Euarctos americanus
Common name: Black Bear – "Smokey"
Sex: Male
Catalog Number: 21,463
Received from: Forest Service and State of New Mexico
How acquired: Gift
Remarks: Weighed 11 pounds upon arrival

The next day, June 30, 1950, the State of New Mexico and the U.S. Forest Service officially presented Smokey to the National Zoo. A crowd of about 500 people attended. The presentation was made by twelve-year-old Stanlee Ann Miller, a granddaughter of Senator Dennis Chavez; eight-year-old Spicer Conant accepted on behalf of the children of Washington.

Smokey was indifferent to the fanfare. After all, it had been a long trip. He lay down and took a nap.

But the media wanted to see Smokey in action, so Pickens put his Forest Service hat on Smokey and Smokey rose to the occasion. Standing on his hind legs, he held the hat in his paws as if he were trying it on. He rolled on the grass and played with the ribbons. The audience was pleased with Smokey's entertainment.

Afterwards, zoo personnel escorted Smokey to his new home. The cage had a large glass window decorated with a Smokey poster captioned "Thanks for helping prevent forest fires." Smokey was not impressed. He shuffled over to the corner and fell asleep. It may be that he dreamed of the pine forest where he had snuggled against his big, furry mother.

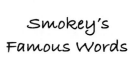

Smokey's Famous Words

23

On the Job

Smokey Bear was the zoo's main attraction—great publicity for the Forest Service fire prevention campaign.

However, Smokey didn't know he was a star, and he mostly ignored the throngs of people who lined up to visit him.

Soon after Smokey's arrival at the zoo, a legendary Hollywood movie star came to town. William Boyd, famous for his role as cowboy hero Hopalong Cassidy, came to Washington with the Cole Brothers circus.

"Wouldn't it be great to get a picture of Smokey with Hopalong Cassidy?" suggested a public relations person. So Smokey went to the circus.

"Let's get a picture of Smokey on my horse," said Hopalong Cassidy.

"Hoppy, you better wear these gloves," said one of Smokey's caretakers. "That bear likes to bite."

Riding a white horse and wearing a fancy white cowboy costume, Hoppy shook his head. "I don't need gloves. He's just a little cub." Then he picked up Smokey by the scruff of his neck and placed Smokey in front of him on the horse. Smokey instantly urinated on Hoppy's white costume. Then Smokey bit Hoppy on the wrist. At this point, Hoppy flung Smokey to the ground.

A circus bareback rider was shocked. "How could Hoppy do that to such a 'cute little bear?'"

Hoppy didn't stay mad at

Hopalong Cassidy and Smokey Bear

24

Smokey. In fact, a year later, he visited Smokey at the zoo wearing the same pants he had worn at their first meeting. In front of Smokey's cage, Hoppy was made an honorary forest ranger.

Another visitor to Washington was Judy Bell. On May 8, 1958, President Dwight Eisenhower made a presentation to Judy Bell on the White House lawn—an "Oscar" trophy. It was awarded to Judy on behalf of the children of America as an appreciation for supporting forest fire prevention. Judy later gave the Golden Smokey award to the Town of Capitan.

After the award ceremony, Judy visited Smokey at the zoo. The press published pictures of Judy and Smokey when Smokey weighed only five pounds and, later, at 300 pounds. Eventually, he reached 500 pounds.

Smokey at 5 pounds

One of Judy's duties, while in Washington, was to help answer some of Smokey's correspondence. In fact, Smokey received so much mail – sometimes 1,000 letters per day – that he was assigned his own personal ZIP code: 20252.

Much of Smokey's mail came from children applying for membership in the Junior Forest Rangers. Each child who wrote to Smokey received a packet of fire prevention materials. Many of Smokey's visitors were proud members of the Rangers. "Look, Smokey, I'm wearing my Junior Forest Ranger badge," said one young visitor. "And, I brought my membership card. I pledge to prevent forest fires," said another.

Smokey at 300 pounds

26

Over six million children became Junior Forest Rangers, and they took their responsibility seriously. One nine-year-old ranger named "Eric" broke the pledge when he started a small fire. Eric mailed back his badge. Smokey replied, "If you do not play with matches and are careful with fires, I will return your badge in three months." Eric accepted the consequences of his behavior and talked to his school about fire prevention. After three months, he wrote to Smokey, "I want my badge back. I haven't played with matches or fire." Eric's mother enclosed a note stating this was true. Smokey mailed Eric a new Junior Forest Ranger badge and a picture of Smokey captioned "Welcome back, Eric."

Smokey goes for a swim.

Here Comes the Bride...

In 1962, the Cooperative Forest Fire Prevention (CFFP) Program asked Ray Bell to find a mate for Smokey. Bell located Goldie Bear in Magdalena, New Mexico. She was an American black bear with gold colored fur, weighing about 150 pounds. Dr. Smith in Santa Fe examined Goldie and pronounced her to be healthy. Goldie was described as being a "nice, practical bear – not mean and nasty."

Bell escorted Goldie to Washington, D.C., where she was taken by motorcade to the National Zoo. There, zoo officials marked the occasion by giving Goldie a wedding ring. However, no little bears were born to this couple. Smokey pretty much ignored Goldie. He preferred to snack on honey sandwiches and swim in the pool.

At the zoo, Bell turned down requests for him to pose with Smokey inside the cage. "Smokey is not a pet. He's a wild bear—500 pounds of wild bear." Bell loved animals, but he knew that wild animals could be dangerous, outdoors or in cages. So he sat outside the cage and watched Smokey. Likely, Bell remembered their adventures when Smokey was a feisty little cub.

After a time, Smokey fans believed Smokey deserved to live in a nicer, more modern cage instead of his old den with stone walls and cement floors. But the zoo didn't have money to build Smokey a new home. So Senator Metcalf of Montana requested a home loan for Smokey from the Housing and Home Finance Agency. The agency replied, "Sorry, we don't loan money to bears." Wanting to help, school children mailed over $4,000 in pennies to build Smokey a new house, but it wasn't enough. Smokey remained in his old, but clean, bear den. He didn't complain. He was, after all, well fed, entertained by his visitors, and even had his own swimming pool.

Later, a plate glass window was installed in Smokey's cage and a concrete tree for climbing and a honey pot were placed inside his yard.

While Smokey lounged around his den, the fire prevention program was going strong. There was a TV special, "Ballad of Smokey Bear," and a Smokey Bear balloon joined the Macy's Day Parade.

"Smokey the Bear" comic strips appeared in newspapers across the nation, and fire prevention advertisements were broadcast over television and radio. Licensed Smokey Bear dolls, books, toys, clothing, and room decorations were sold.

Across the United States, children slept on Smokey Bear sheets, ate Smokey Bear cereal, bathed with Smokey Bear soap, and sang "Smokey the Bear."

These were busy years for the Smokey Bear fire prevention campaign.

Senior Years

After 25 years (comparable to 70 human years), Smokey had health problems. He suffered from arthritis and had difficulty walking. But he was not forgotten. A family in Biddeford, Maine, read in the newspaper about Smokey's poor health. They sent him a letter of sympathy and a pound of honey.

The zoo keepers were concerned. "It may be time for Smokey to retire. He spends more time sleeping and rarely comes out to see his fans."

Smokey Bear Balloon in Macy's Day Parade.

Retirement

On May 2, 1975, the National Zoo hosted a retirement party for Smokey. Government officials and dignitaries attended the ceremony and a spokesman for the Forest Service made the announcement, "Ladies and gentlemen, boys and girls, Smokey Bear is the first bear to become a member of the National Association of Retired Federal Employees." Hundreds of people applauded.

Rudy Wendelin, the Smokey Bear artist, painted a picture for the occasion with the caption "Carry On, Little Smokey." The picture showed an older ragged-looking Smokey handing his shovel and pants over to Little Smokey, a wide-eyed cub holding a ranger hat in his paws. At Smokey's feet was a well-used suitcase with New Mexico travel stickers, the dates 1950 and 1975, and autographs of friends and famous people.

The Cooperative Forest Five Prevention Program also presented a poster for the occasion.

Smokey Bear had had a successful career. He and the Smokey Bear campaign had been responsible for reducing human-caused forest fires by 70%. Smokey had earned over one million dollars in

Retirement poster drawn by Rudolph Wendelin

31

THANKS FOR LISTENING.

Cooperative Forest Fire Prevention Program retirement poster.

royalties from licensed products and was well on his way to the next million. He deserved the recognition.

Then came a sad announcement. "The title of Smokey Bear is awarded to Smokey II. He came from the same forest where Smokey was found. He also lost his mother." Smokey II, a scrawny little cub weighing about 40 pounds, was brought to the platform. The audience clapped. He was then ushered to the living quarters reserved for Smokey Bear.

Smokey and Goldie were moved to a common cage on Bear Row. Now, the plaque over Smokey's cage simply read, "American Black Bear."

The Final Flight

On November 9, 1976, a quiet cloud settled over the zoo. Goldie alerted workers that something was wrong by moaning and swaying back and forth. The workers found Smokey's body. In hushed voices, they passed along the sad news, "Smokey died in his sleep."

Prior arrangements had been made to return Smokey to New Mexico for burial. He was packed into a crate and loaded into the cargo hold of TWA Flight 217. Smokey was going home.

The pilot, Raymond Lutz, was from Taos, New Mexico. He announced the flight time, flying altitude, and usual details. He then added, "This flight has special cargo – the remains of Smokey Bear."

When they landed in Albuquerque, they were met by TV cameras, the New Mexico State Police, and Forest Service officials. There were rumors that Smokey might be hijacked, so Smokey's coffin was quickly loaded onto the back of a Forest Service pickup. The State Police escorted Smokey from Albuquerque to Capitan. To keep Smokey's burial site a secret, he was buried at night by the headlights of a few vehicles.

Smokey Bear's grave stone and gravesite waterfall. 33
Photo taken by Sue Houser

Farewell, Smokey Bear

A dedication service was held on November 17, 1976, at the Smokey Bear Historical State Park in Capitan, New Mexico. About 250 people attended. Flowers and funeral wreaths filled the Smokey Bear Museum and lined the porch and sidewalks. Children sent letters expressing their sympathy. From Tucson, Arizona, one letter read,

"Dear Mr. Forest Ranger.
I read a story in the newspaper
about Smokey and how he died.
It made me feel sad."

The service began with a prayer by a local minister. Two of Smokey's early friends, Elliott Barker and Ray Bell, talked about Smokey Bear—the living symbol that had been born as a poster bear. The poster bear had shared his name with the live American black bear cub and together, they had waged an ambitious, successful fire prevention campaign.

In a letter from Rhode Island:

". . .It would have been great to spend even a
minute with Smokey. Sure, he wouldn't understand
me if I talked to him; I probably wouldn't have
gotten to see him and him to see me-just that instance,
no matter how brief, would have suited me. And I've
had all my 15 years to do that. And how I wanted to.
Now it's too late. . ."

The living symbol of fire prevention was gone. However, Smokey Bear, the poster bear, lives on, continuing to remind the public of wildfire safety and fire prevention.

His message had survived as shown in a letter written by a child from New York :

"Dear Smokey,
I'm sorry that you died . . . We will
all miss you an will be very careful
with fires."

34

AFTERWORD

Today, public awareness of Smokey Bear remains high, but wildfire awareness is low. Because of a dramatic increase in wildfires (8.4 million acres in 2000; 6.9 million acres in 2002), Smokey Bear's fire prevention message is more vital today than ever before.

In 2001, a new public service advertising campaign was produced by the Advertising Council, the USDA Forest Service, and the National Association of State Foresters. This campaign changed the tone and audience of Smokey's message. While the campaign remained focused on children, it also embraced the need to educate adults. A new tagline, "Only You Can Prevent Wildfires," was adopted and, in 2002, a new website created.

Smokeybear.com features an informational, adult-directed site called "Only You" that provides factual and scientific information to individuals who frequent America's forests. Its redesigned children's area includes Smokejumpers!, an online game that honors the firefighters who parachute out of planes. "Smokey's Vault" is a collection of posters, images, and public service announcements for nostalgic Smokey fans. This website was created by the Ad Council and Ruder Finn, Inc., an international communications agency that impacts public opinion and affects change.

The famous Smokey Bear symbol and message still appear on radio, television, billboards, and in newspapers. Websites regarding Smokey Bear are: Smokeybear.com and symbols.gov.

Remember, "Only You Can Prevent Wildfires."

APPENDIX A
IN THE NAME OF SMOKEY BEAR

SMOKEY BEAR LAW – On May 23, 1952, Public Law 359 of the 82nd Congress was passed to protect Smokey's picture or name from being used in the wrong way. Amended in 1974 as Public Law 93-318 of the 93rd Congress, it is better known as the "Smokey Bear Act." Thus, his official name is Smokey Bear, not Smokey the Bear.

SMOKEY THE BEAR SONG* – This famous song was written by Steve Nelson and Jack Rollins, under license of the U.S. Dept. of Agriculture to Hill and Range Songs, Inc., and recorded under license by several recording companies, © 1952. Words to the familiar chorus are:

Smokey the Bear, Smokey the Bear.
Prowlin' and a growlin' and a sniffin' the air.
He can find a fire before it starts to flame.
That's why they call him Smokey,
That was how he got his name.

*The composers were allowed to add "the" to the song for the rhythm of the score.

SMOKEY THE BEAR RECORD – Ranking fifth place in record sales during the 1952 Christmas season, this song was recorded by Victor, Columbia, Golden and Decca Records.

SMOKEY BEAR TOY BEAR – In 1952, Ideal Toys manufactured the first Smokey Bear stuffed animal for sale. Each stuffed bear was sold with an application card to fill out and mail to the Smokey Bear Headquarters to become a "Junior Forest Ranger." Upon receipt of the completed application card, a kit containing a membership card, badge and other items were sent to the child. A reissue of the Smokey Bear Toy Bear was initiated in 2004.

JUNIOR FOREST RANGER PROGRAM – Before Smokey Bear's death, over six million children enrolled as members of the Junior Forest Ranger Program. Any child interested in becoming a member can write to:

Smokey Bear
USDA Forest Service
402 SE 11th St.
Grand Rapids, MN 55744

SMOKEY BEAR'S ZIP CODE – By 1964, Smokey was receiving more mail than even the President of the United States. On April 29, 1964, Deputy Postmaster William M. McMillan assigned a personal ZIP code 20252 to Smokey Bear.

SMOKEY THE BEAR COMIC BOOKS AND CARTOONS – Between the years 1955 and 1973, more than two dozen Smokey the Bear comic books were published and a daily comic strip appeared in newspapers nationwide from 1957 to 1960.

Paul S. Newman wrote the majority of the Smokey the Bear comics which were illustrated by Morris Gollub. Comic strips were also written and illustrated by Newman and Gollub under the byline "Wes Woods."

SMOKEY BEAR BALLOON – In 1966, the Macy's Thanksgiving Day Parade in New York City introduced a Smokey Bear balloon that was 59 feet tall. It was in the parade several times after that, including the 1993 Thanksgiving Day Parade when Smokey celebrated his upcoming 50th birthday. The balloon has also appeared in the Tournament of Roses Parade in Pasadena, California.

THE BALLAD OF SMOKEY THE BEAR – This television show premiered on Thanksgiving Day, November 24, 1966, on the General Electric Full Color Fantasy Hour. This animated Rankin/Bass Special was written by Joseph Schrank and narrated by James Cagney. The composer was Johnny Marks and Maury Laws wrote the musical score.

SMOKEY BEAR STAMP – For Smokey's 40th birthday, Rudy Wendelin created the art for a commemorative stamp. It shows the Smokey Bear advertising symbol in the background with the live bear cub clinging to a burned brown snag in the foreground. It was issued by the U.S. Postal Service in Capitan, New Mexico, on August 13, 1984.

FRIENDS OF SMOKEY BEAR BALLOON – Bill Chapel, a hot air balloon pilot, and others designed the Smokey Bear Hot Air Balloon. Almost 100 feet high and weighing 1,100 pounds, the Smokey Bear balloon made its debut before one million visitors at the 1993 Albuquerque, New Mexico, Balloon Fiesta.

The Smokey Bear Balloon suffered a tragic mishap at the Albuquerque International Balloon Fiesta on October 11, 2004, when crosswinds pushed the balloon into a 50,000 watt radio tower. The pilot, Bill Chapel, and passengers, ten-year-old Aaron Whitacre and 14 – year-old Troy Wells, inched their way down the 670 foot tall swaying tower. Employees of the Public Service Co. of New Mexico and local firemen assisted the three brave ballooners who safely reached the ground.

Friends of Smokey Bear Balloon immediately rallied and raised funds to acquire a new Smokey Bear balloon – a duplicate of the original. The balloon was picked up in August, 2005, and test flights were made in the Black Hills National Forest at Sturges, South Dakota, and other locations. The Smokey Bear balloon flew among the thousand hot air balloons at the October 2005 Albuquerque International Balloon Fiesta, once again piloted by Bill Chapel.

Friends of Smokey Bear Balloon, a non-profit corporation, also has four cold air, static display balloons that are used for promotional purposes to protect America's natural resources.

SMOKEY BEAR MUSEUM – The log cabin museum in Capitan, New Mexico, opened its doors in 1961. It is overflowing with Smokey memorabilia, photos and posters as well as items related to Smokey's business: Milton Bradley's Smokey Bear game, Smokey canteens, books, hats, dolls, watches, pens, scarves, rulers, Viewmaster reels, Smokey mugs, patches, banks, shovels, ashtrays, coins, stamps, comics, and sweatshirts.

SMOKEY BEAR HISTORICAL STATE PARK – On May 15, 1976, the Smokey Bear Historical State Park was dedicated. At the time, it was a dirt lot with a lone sign. Dedication ceremonies reflected the spirit of Capitan's favorite bear and the fire prevention program. Here, in the corner of the park, Smokey Bear was buried on November 9, 1976. The park is now shaded by trees. A winding walkway leads to a giant boulder bearing a bronze plaque that pays tribute to Smokey Bear. A waterfall gurgles nearby.

SMOKEY BEAR STAMPEDE AND RODEO – This event is held in annually on the Fourth of July in Capitan, New Mexico.

Smokey Bear Historical Park, Capitan, New Mexico.
Photo taken by Sue Houser

Capitan is located in a valley surrounded by the Lincoln National Forest on US 380 in southern New Mexico. If you go to Capitan, you might want to stay at the SMOKEY BEAR MOTEL AND CAFÉ located on SMOKEY BEAR BOULEVARD.

HOT FOOT TEDDY COLLECTOR'S ASSOCIATION – In 1994, Jim VanMeter, a retired California forest ranger, founded the association for collectors of Smokey Bear memorabilia. This 400 strong membership hails from coast to coast and border to border with extensive collections of photos, dolls, ceramics, posters, and many more items. They hold regional meetings and national conferences and publish a quarterly newsletter. Anyone interested in membership may contact President Jan Fite at 505-445-2039 or email: marvinjan6@msn.com or at www. hotfootteddy.org..

APPENDIX B

HAPPY BIRTHDAY, SMOKEY BEAR!

August 9, 1944, is Smokey Bear's official birth date. That is when the "idea" of Smokey Bear was born. To spread the message about forest fire prevention, the Cooperative Forest Fire Prevention (CFFP) Program, chose a bear – a strong, powerful forest dweller.

It was established that the advertising bear was to be black or brown. His face should look like a panda bear's face, and have a quizzical expression. And he should wear a hat. (The pants were added later.)

Although the live Smokey Bear cub was rescued in 1950, his birthday is celebrated at the same time as the advertising bear. However, when the live Smokey turned twenty years old on March 31, 1970, he was remembered by his friends in Santa Fe, New Mexico. The Santa Fe Wildlife and Conservation Association put on a party, and the Department of Game and Fish sent a birthday cake to Smokey at the zoo in Washington, D.C.

OFFICIAL CELEBRATIONS

SMOKEY'S 30TH BIRTHDAY

In 1974, the advertising campaign created five "THINK" messages for Smokey. The word "THINK" was written in huge letters with a fire prevention message underneath. The letter "I" in the word "THINK" was replaced with a forest symbol, for example: a picture of Smokey, Bambi, or a tree.

SMOKEY'S 40TH BIRTHDAY

In honor of Smokey's 40th birthday, Rudy Wendelin designed a postage stamp that was issued by the US Postal Service on August 13, 1984, in Capitan, New Mexico. At the time, a first class stamp sold for twenty cents.

The USDA Forest Service unveiled a collection of posters representing Smokey's first 40 years.

The North Carolina Forestry Division produced a 21-foot tall animated Smokey that traveled all over the state. Ten years later, the animated bear traveled to Washington to attend Smokey Bear's 50th birthday party.

SMOKEY'S 50TH BIRTHDAY

Smokey had so many friends all over America and other countries that it took one entire year to celebrate his 50th birthday. The U.S. Forest Service issued a video presentation about Smokey Bear. Traveling museums carried Smokey Bear exhibits and his history.

Then, on August 9, 1994, an outdoor event was held on the Ellipse, the park south of the White House in Washington, D.C., with music, entertainment, exhibits and a person wearing a Smokey Bear costume. When fans on the Ellipse saw the Smokey Bear mascot on live television leading a parade at Disney World in Florida, the crowd joined in singing "Happy Birthday" to Smokey.

SMOKEY'S 58Th BIRTHDAY

For his birthday, Smokey was given his own website: www.Smokeybear.com. The sponsors for the website are the USDA Forest Service, National Association of State Foresters, and the Advertising Council.

Ruder Finn, Inc., international experts in design and web technology, and the Advertising Council, Inc., a private, nonprofit organization producing public service advertising campaigns, launched the new Smokeybear.com website on August 9, 2002. This updated site is directed toward adults as well as children, including Smokejumpers!, an online game honoring firefighters who parachute out of planes to fight wildfires.

SMOKEY'S 60th BIRTHDAY

On May 7-9, 2004, Capitan, New Mexico, overflowed with people. Some came from as far as Washington, D.C., to celebrate Smokey's 60th birthday. At the Smokey Bear State Park, the Smokey Bear balloon swayed to the music of the Longhorn Dance Band and local musicians.

Horse drawn carriages ferried people to activities around the bustling little town. There were horseshoe tournaments, chainsaw bear carving competitions, an antique and custom car show, a Smokey Bear play, puppet shows, and memorabilia exhibitions as well as local arts and crafts. The Highway 101 Concert opened with a performance of Jim Raby and the Desert Stars.

Foote, Cone & Belding, the advertising agency that has worked with the Smokey Bear campaign for many years, designed a logo for the occasion – a picture of Smokey encircled with the words "Sixty Years of Vigilance." This logo and Smokey's 60th birthday cake can be seen on the Smokeybear.com website.

Smokey Bear was recognized by Congress on August 9, 2004, for 60 years of fire prevention campaigns.

A celebration was held in Disneyland to honor Smokey on his 60th birthday and to mark the renewed partnership with Bambi and Disney.

Celebrating Smokey's 30th
Birthday

Smokey Bear
"Sixty Years of Vigilance" Logo.
Courtesy of Smokeybear.com website.

SIXTY YEARS OF VIGILANCE

SMOKEY

39

Bibliography

"All in a Day Smokey Bear Historical State Park," New Mexico Magazine, January, 1993.

Barker, Elliott S., "Smokey Bear and the Great Wilderness," Sunstone Press, 1982.

"Happy Birthday, Smokey," New York State Conservationist, August, 2004.

Higgins, Philip, "Ere Smokey Departs, Say Foresters, He's Got Party to Attend," Santa Fe New Mexican, June 25, 1950.

Higgins, Philip, "Smokey Heads to Washington," Santa Fe New Mexican, June 27, 1950.

"I'm a Stranger Here Myself," Santa Fe New Mexican, May 14, 1950.

Lawter, Jr., William Clifford. "Smokey Bear 20252, A Biography," Lindsay Smith, 1994.

Morrison, Ellen Earnhardt. "Guardian of the Forest," Morielle, 1976.

Morrison, Ellen Earnhardt. "The Smokey Bear Story," Morielle, 1995.

Pickens, Homer C., "Tracks Across New Mexico," Bishop Publishing Co., 1980.

"Smokey Bear Hits Tower; 3 Escape Injury," Albuquerque Journal, October 11, 2004.

"Teddy With A Hotfoot," Santa Fe New Mexican, May 11, 1950.

"Tidbits," American Profile, Insert in Albuquerque Journal, January 20, 2004.

"The Saga of Smokey Bear," New Mexico Magazine, March, 1984.

Other Sources

Bell, Don; Las Cruces, New Mexico, original interview.

Bervin, Dave; New Mexico State Forestry, Bernalillo, New Mexico, Photograph of Smokey Bear.

Fite, Jan; President of Hot Foot Teddy Collector's Association, Raton, New Mexico, original interview.

Griner, Carol and Dallas; Friends of Smokey Bear Balloon; Albuquerque, New Mexico, original interview and photographs.

Rouse, Robert; Las Vegas, Nevada; Former President of Hot Foot Teddy Collector's Assoc., original interview.

Smokey Bear Museum, Capitan, New Mexico, photographs and children's letters.